Voices for Vermont Yankee

Meredith and George Angwin

Carnot Communications
www.carnotcommunications.com

Published March 2013 by Carnot Communications.
Printed in the United States of America
ISBN 978-0-9891190-0-9

★★★★★

The statements in this book were supplied to Meredith Angwin by their authors, from whom permission to print was obtained. Similar material can be found in the public domain because statements for Public Service Board meetings are entered into the public record. The public record statements have not been compared to those in this book, so no claim is made that the two versions are exactly the same. To review the public record, contact the Vermont Public Service Board referring to docket 7862.

All other text in this book was written by Meredith and George Angwin.

The photograph of Vermont Yankee on the cover and title page is by the Nuclear Regulatory Commission and is in the public domain.

The photographs of Carl Pinkham, Lindsay Rose, and Meredith Angwin were taken by Entergy staff. These are copyright 2012 by Entergy Nuclear Vermont Yankee and used by permission.

The photo of William Schulze was taken by Howard Shaffer, and the photos of Deb Shulze and Stuart Endsley were supplied by Shulze and Endsley, respectively; all used by permission.

The photograph of the button "Yes Vermont Yankee" was taken by Meredith Angwin.

The remaining photographs were taken by Stephanie Thomson of Vermont Energy Partnership. They are copyright 2012 by Vermont Energy Partnership and used by permission.

Meredith and George Angwin give grateful thanks to those who provided statements and photographs for this book.

Table of Contents

Introduction

This book contains twenty-one statements from ordinary citizens who support the continued operation of the Vermont Yankee nuclear power plant. These statements describe Vermont Yankee as valuable, clean, safe, and supporting its community.

People who think there are substantial benefits to using nuclear power have had little visibility in newspapers, magazines, and broadcast media. Opponents of nuclear power have dominated these news and opinion venues. While many articles discuss potential dangers of the plant, few articles discuss its positive impacts on the environment, especially compared to alternatives. We have produced this book to correct the one-sided treatment in the media and to provide a resource for Vermont Yankee supporters.

Public Service Board Hearings

In order for a nuclear power plant to operate, it must have a license from the Nuclear Regulatory Commission and authorization from the state in which it operates. The NRC is responsible for regulating nuclear safety in the United States. Vermont's Public Service Board is responsible for issuing the state authorization, called a Certificate of Public Good.

The NRC has issued a license renewal to Vermont Yankee, but the plant still needs state authorization. Although there are legal challenges still in the court system, Vermont's Public Service Board has elected to proceed with hearings on issuing a certificate.

Vermont's Public Service Board opened Docket No. 7862 for the Certificate of Public Good. It held two hearings for the public to comment on issuing the certificate. These were the only hearings where the public was invited to speak. Other hearings will be held in which people with official standing – lawyers, experts, and intervenors – will present testimony. Although there are no more public hearings planned, you may submit a comment on this docket through the Public Service Board's web page.

The first hearing was held on November 7, 2012, in the Vernon Elementary School across the road from the entrance to Vermont Yankee. People lined up to give short speeches in favor of or against issuing the certificate. As you can see in this panoramic photograph, the hearing attracted a big crowd.

The second hearing, on November 19, 2012, was held by interactive TV in several locations around Vermont. At that hearing, people gave their statements in a TV studio arranged so they could be seen by members of the Public Service Board and all the attendees.

Publishing the Statements

At the hearings, the breadth and depth of the statements made by plant supporters impressed Meredith. She asked some of the supporters for permission to post their statements on her blog *Yes Vermont Yankee*. She collected twenty statements from the hearings, plus one submitted by email, and published them as guest posts.

Meanwhile, Meredith was becoming aware of the increasing importance of ebooks. She asked George, her husband and colleague in Carnot Communications, to look into making an ebook of the blog posts. George returned with a favorable report. He organized the guest posts to tell a story, created front matter, added photographs, and published the ebook *Voices for Vermont Yankee*. It is available for the Kindle from Amazon and for the Nook from Barnes & Noble. It can be read on other platforms via free downloadable apps.

Within days of the publication of the ebook, Meredith got requests for printed versions. So George studied ways to produce a professional-looking printed book. This task was more difficult than the ebook in two ways: a printed book must satisfy stronger formatting requirements, and one must produce actual physical objects and distribute them. The text for the printed book is the ebook text, but, in a few cases, Meredith or George made small corrections for clarity while preserving the authors' characteristic voice. You are now reading the result of this investigation and work.

Purpose of This Book

We expect that the viewpoints expressed in this book will provide a more complete and compelling case for the continued operation of Vermont Yankee than was previously available. We also hope that it may inspire others to submit statements by email to Vermont's Public Service Board or perhaps to take part in similar hearings in other parts of the country. Finally, we hope that our book will help people all across America appreciate the benefits that nuclear power has given us and can give us in the future.

Valuable to Vermont

1. A Strong Vision for the Future of Vermont

By Charles Kelly

Reference PSB Docket No. 7862.

Gentlemen of the Vermont Public Service Board,

I am sure you have heard all the arguments before but some things are worth repeating.

The electrical generating capacity of Vermont Yankee needs to be continued as well as Vermont's use of it. As you know there is no Vermont based generated energy on-line or feasible in the next decade that can replace Vermont Yankee's near 5,000 giga-Wh per year. Vermonters do not discard useful tools before their time. Please do not discard this valuable in-place, environmentally clean, energy source.

Reliance on foreign and out-of-state energy sources, coupled with our own meager intermittent renewable sources, leaves us vulnerable to uncontrolled rate increases and in a weak bargaining position.

At some time in the distant future, a replacement state-of-the art energy system can make valuable use of the pre-qualified central generating site in Vernon.

Let the legacy of the 2012 Public Service Board be one of vision:

- Vision to see through the political rhetoric.
- Vision to not be fooled by the profiteering antics of the intermittent energy coalition.
- Vision to provide affordable energy for Vermont's working class and small business community.
- And a vision to keep Vermont Yankee generating for Vermont.

Thank you for your attention.

Sincerely,
Charles Kelly
Addison, Vermont

★★★★★

Charles Kelly drove from Addison, Vermont, to speak at the Vernon hearing. He holds BME and MME degrees from Villanova University and is a Professional Engineer in Pennsylvania and Vermont. You can read more about Kelly's background at his guest post "PV Solar and Vermont, Not a Good Fit" on the blog *Yes Vermont Yankee*. Kelly serves on the planning commission for the Town of Addison.

2. Power, Carbon and Costs

By Peter Lothes

Petition Urging Granting a Certificate of Public Good to Vermont Yankee

My name is Peter Lothes. I live in Quechee Vermont, and I am here to urge the granting of a Certificate of Public Good to Vermont Yankee so it may continue to operate. As a retired engineer and business manager, I believe closing Vermont Yankee would result in a terribly negative impact upon the state's economy, the well being of its residents, and our mutual goal of reducing carbon in the atmosphere.

We need Vermont Yankee for a number of reasons:

First: We need the power capacity

Over time, the need for power will grow in Vermont, and the rest of the northeast grid which Vermont Yankee supports. Without Vermont Yankee, the additional power will have to come from new construction of fossil plants. This will be costly, and will increase carbon emissions. Obtaining power from Hydro Quebec (which itself includes nuclear power) may seem in the short term to solve Vermont's problem, but buying their power does nothing for the rest of the area that Vermont Yankee serves, and relying on a power source from outside of the US has its own very considerable risks. Every knowledgeable expert I have read agrees that other low/no carbon generators (wind, solar, thermal etc.) cannot provide the base load of electrical power that will be needed in the future. Dr. James Hansen of the Columbia Earth Institute (one the world's most respected environmentalists) makes this case persuasively in his *Storms of my Grandchildren* among other writings.

Second: We need it to contribute to the world's control of carbon emission.

Dr. Hansen and others (e.g. Dr. Robert Hargraves, Dartmouth Professor, PhD physics) argue persuasively that only the development of nuclear power for base load, can keep China, India, and the United States from burning the huge coal resources each of these countries has. This is because only nuclear power can generate electricity cheaper, significantly cheaper, than coal. They will all burn the cheapest fuel, and if they do so, nothing else that the rest of us do will matter, and carbon emissions will grow dramatically.

Third: We need it to maintain our expertise in the nuclear power technology

We have, in Vermont Yankee, solid nuclear power technology and experience, upon which we can build in the future, as nuclear power becomes an important industry. This is an industry which can grow dramatically, providing technology driven manufacturing, scientific , clerical and administrative businesses and employment. It could be a significant export opportunity for the US, and will provide jobs and economic growth that would not be outsourced. In the 70's's and 80's, we lost our nuclear technology leadership base, and now the next generation of power reactors are being built in France and China. They ultimately will generate electricity at about 4 to 5 cents per kWh. If we continue to lose this technology, we will be importing, in addition to plastic toys and clothing, nuclear power plants built in China.

The nuclear industry will have good paying, reliable businesses and jobs, and we need to grow them here in Vermont.

Lastly: We need it to contain the cost of electricity in this state to keep and attract new businesses, and citizens.

Vermont is a very unfriendly cost environment for businesses and individuals, largely due to our tax and distribution structure. Electrical power cost has been one of the few areas where the price to consumers, while not advantageous, is not excessive. It has been kept that way in the past due to the very low cost of Vermont Yankee power.

I understand the following to be the costs distribution utilities here now pay for power from various sources (data from memory, and could be off somewhat):

- Solar: 27 cents per kwh
- Wind 15 cents
- Vermont Yankee 6.1 cents, plus revenue sharing
- Quebec Hydro: 5.9 cents
- Coal: 12 to 13 cents (in China 6 to 8)

The wrong selection of power sources would drive our electricity costs ever higher, and drive businesses like GE and IBM out of the state. It will make it impossible to attract new industry, or even attract retirees. They are a considerable asset to the state, and they struggle under the tax burden.

If we fail to participate in this energy transition, our Vermont economy will stagnate or worse, and we will suffer significantly more carbon emissions than would be the case with Vermont Yankee operating.

Respectfully,
Peter B. Lothes
Quechee, VT 05059

3. Baseball, Reliable Power, and Vermont Yankee

By Richard Trudell

Good evening, my name is Dick Trudell, I live in Grand Isle. I'm a civil engineer with professional licenses in Vermont and Massachusetts. I founded Trudell Consulting Engineers in Williston, employing upwards of 15 people and am now retired.

Over a career spanning 41 years, I've learned not to take on projects you don't have the ability or technical capacity to complete. Being a native Vermonter, I also feel that I have a certain amount of common sense. Why do these two things compel me to drive 360 miles round trip to spend a couple of minutes testifying to this board?

I'll give you an analogy.

Vermont Yankee has proved to be dependable source of baseload power for Vermont with approximately half of its 620 MW capacity serving the homes and businesses of Vermont, with over an 80% capacity factor. Now if you had someone on your team batting 0.800, it is unlikely you would kick them off the team – that's just plain common sense. Some Vermonters still seem to think that with enough conservation, plus solar and wind power, we can replace Vermont Yankee's 620 baseload megawatts with a couple of rookies that are batting at best 0.300, require state subsidies before they will even go to the locker room to suit up, and their salaries cost more that the dependable pro you have had for years. Then you find out they can only play for a couple of years before they need surgery or knee replacements. You have got to be kidding me.

You know that Vermont Yankee makes on-demand power, but solar and wind are "intermittent." That is why they bat only 0.300. With our current technological abilities, too many 0.300 players will give you a bad season, and too high a proportion of intermittent power will challenge, and could eventually crash the transmission grid.

The State of Vermont's plan acknowledges this shortcoming, but its solution – building biomass and natural gas plants and buying lots of out of state nuclear, coal and especially natural gas power – has its own permitting, environmental, and cost problems, that to me exceed the cost of maintaining and improving Vermont Yankee, and, if I may add, eventually replacing the existing boiling water reactor plant with a state of the art breeder reactor plant at the same site, hence reducing or eliminating the problem of spent fuel storage.

In my opinion, we can build renewables gradually and sensibly only if we keep existing baseload sources, especially those like Vermont Yankee, which are emissions free, reliable, and provide outstanding economic benefits. And those renewable resources have to be cost effective as well, operating without subsidies. Until some engineer, more clever than me, finds a way to, in effect, turn intermittent power into baseload, this seems like the prudent way to go.

Vermonters are right to want clean air. We are right to want energy independence. We are right to want to keep what is left of our manufacturing base. For all of these reasons, we are right to ask you to keep Vermont Yankee operating.

Thank you
Richard Trudell

4. Affordable Reliable Electricity

By Dianne Amme

Nuclear power is reliable and affordable and thus essential for quality of life.

I am personally dismayed that as a Vermont resident I don't get to purchase electricity from Vermont Yankee, but thank goodness the plant continues to operate and provide safe, carbon free, reliable affordable power for some lucky customers as well as provide many jobs.

Please approve the Certificate of Public Good.

★★★★★

Dianne Amme is a dental hygienist living in Norwich, Vermont. She has three sons.

5. Farm and Forest in Vermont

By Bruce Shields

I am Bruce Shields, President of Ethan Allen Institute, a Vermont organization promoting free markets and individual decision making. I have been active in agriculture and forestry in Vermont since 1977.

Agriculture and forestry at about 20% of Vermont's economy provide a stable and widely distributed base of employment. Thousands more derive their living indirectly -- paper mills, Cabot Cheese, Ethan Allen Furniture and hundreds of self employed truckers, mechanics, and financial service workers.

These people look to State government to provide a stable and predictable economic framework in which they may operate. Energy is key to the success of farming and forestry. No other sector of Vermont has more rapidly and profoundly modernized its production process, and electrical energy is their indispensable base. One modern farmer with several hundred horsepower of electrical motors on his farm creates more value than 50 farmers like my grandfather did. An electric sawmill with 5 or 6 hands can put up as much lumber in a day as 30 or more hands could 50 years ago.

Both farming and forestry are dominated by small family businesses who have little influence over their selling price. The market provides our income: we can only control our expenses. Electricity is one of the largest expenses for the typical farm. In 2009, Vermont Gross Farm Income was $589,827,000. The National Agricultural Statistical Service [NASS] shows that electricity cost our farms $17,423,000 -- 2.95% of GFI. After all expenses are deducted, Net Farm Income for 2009 was $97,099,000. Electricity cost equals 17.95% of NFI. Both the low net operating margin, and the high cost of electricity are consistent with prior years. But the trend is adverse: in 2003, electricity was 2.72% of GFI, and just 11.87% of NFI. That upward trend in electrical costs is a negative for business viability.

Vermont Yankee has anchored the power costs of all New England. Steady and predictable generation smooths price volatility across the region, helping to cap spikes and fill in voids. Other power sources may move in concert with dominant economic indicators and thus impact farmers and sawmills concurrently with other stressors like high oil costs, or a drought. For Vermont's economic health we need to maintain as many counter-cyclical mechanisms as possible.

An operational Vermont Yankee will also inject about $100 million per year into the Vermont economy, and pay millions in statewide education taxes which already fall heavily on farm and forest land. If Vermont Yankee remains open, some of that $100 million in economic benefit will find its way to the wood products makers and farmers. But if it closes, a share of those millions in lost taxes would be inevitably transferred to them, as well. I see no upside to closing Vermont Yankee for farmers and wood products manufacturers.

★★★★★

Bruce Shields is past president of the Ethan Allen Institute. Meredith's Energy Education Project is part of the Institute. This is a quote from his biography on the Institute website:

> *Bruce, a family tree farmer, has been active in farm and forestry work for many years. He has served as state treasurer of the Vermont Farm Bureau since 1994 and as vice president and president of the Lamoille County Farm Bureau for 13 years. He has been active for many years in leadership positions in the Vermont Timber Truckers and Producers Association, the Vermont Sugar Makers Association, the Vermont Forestry Foundation, and the Vermont Forest Products Association. (Bruce is on the board of directors of the Vermont Forest Products Association.)*

6. Just Plain Good Sense to Issue the Certificate

By William H. Schulze

Members of the Public Service Board, good evening:

My name is William Schulze, I'm a employee of Entergy/Vermont Yankee.

Nearly 33 years ago, I decided that my future and the future of my family would be spent in the Green Mountain State and came to work for the Vermont Yankee Nuclear Power Corporation.

I'm one of those "650 good paying (Vermont Yankee) jobs" that you've heard tell about.

I started at an entry level position in the Operations Department, and after an intense training period, received my NRC Reactor Operator's License then (after an even more intense period of training) my NRC Senior Reactor Operator's License.

For the last 16 years, I've been training operators to obtain and then to maintain their own NRC licenses.

That's where I'm coming from...

This Certificate of Public Good, as I understand it, it is a device that the state of Vermont uses in order to ensure that a business's operation will be both beneficial to people of the state of Vermont and at the same time not detrimental (a pretty reasonable method to look out for the continued welfare of the folks who live here).

Over the years, Vermont Yankee has petitioned for and received these certificates from the PSB for changes in the facility such things as power up-rate, and spent fuel storage.

Vermont Yankee is one of the main reasons why Vermont has the second-lowest per-capita carbon footprint of any state in the U.S. Since 1972, operating Vermont Yankee has prevented 50 million tons of carbon and other pollutants from being released into the environment.

Since 2002, Vermont Yankee has saved its New England customers approximately $330 million in electricity rate savings compared to the price of purchasing that same power from the electrical spot market.

In 2011, Entergy Vermont Yankee contributed approximately $435 thousand dollars to the community through Entergy open grants, the Entergy Charitable Foundation, site sponsorships, annual events.

Also, Vermont Yankee and its employees have provided millions of dollars in economic benefits to the Vermont economy, through payroll, state and local taxes, and the purchase of goods and services from local businesses.

From my perspective, looking at the all the environmental and economic positives that continued operation of the Vermont Yankee facility would present to both the New England region as well as the people of Vermont, it makes just plain good sense to issue Entergy Vermont Yankee a certificate of Public Good.

I urge you to approve and issue it.

Thank you for your kind consideration.

Clean Electricity

7. Air Pollution and Vermont Yankee

By Meredith Angwin

My name is Meredith Joan Angwin and I live in Wilder Vermont. I am here to speak in favor of granting Vermont Yankee a Certificate of Public Good for continued operation. I am the Director of the Energy Education Project of the Ethan Allen Institute, I blog at *Yes Vermont Yankee*.

I am a physical chemist by training. I worked at improving pollution control methods and corrosion resistance of nuclear, gas, geothermal, and coal plants. I was a project manager at the Electric Power Research Institute. I also consulted with many utilities, in the U. S. and abroad.

I am here today as a citizen of Vermont who wants Vermont to remain the clean, green and attractive state that it is today. Nuclear power has the least environmental impact of all baseload types of electricity. Specifically, it creates no air pollution. Nuclear power creates no nitrogen oxides.

Intermittent renewables like solar and wind must be backed up by baseload power and dispatchable power. What kind of backup power will Vermont choose? Hydro, nuclear or fossil?

New hydro plants and new nuclear plants are unlikely to come on-line in this region. Our practical choices are Vermont Yankee, new fossil plants, or buying power from outside Vermont. I will discuss the environmental issues of natural gas versus Vermont Yankee, because I have technical expertise in this area.

Fossil power means air pollution. Natural gas plants are the best in terms of emissions, but they emit acid gases to the air: carbon dioxide and nitrogen oxides. I will talk about nitrogen oxides, an acid gas that contributes to acid rain and smog. I have patents in the control of nitrogen oxides.

Controlling nitrogen oxides is difficult. At the high temperatures in gas turbines, the air actually burns itself. That is, the nitrogen in the air combines with oxygen in the air and makes nitrogen oxides (NOX). NOX is only partially controlled by ammonia addition at the end of the process. Sometimes the ammonia itself becomes a pollutant.

NOX is a very acid gas, contributing to acid rain. NOX is also the main cause of smog, which can happen on any sunny day. You don't have to be in Los Angeles to get smog. All you need is NOX and sunshine.

Nuclear plants do not release NOX. They keep our air clean. For clean Vermont air, we need to make our baseload power with Vermont Yankee, not fossil fuels.

8. Acid and Air

The Environmental Effects of Non-Nuclear Electricity

By Stuart Endsley

My name is Stuart Endsley and I am in favor of granting a permit to Vermont Yankee. Over the past 32 years, I have worked in many types of power generation facilities — both nuclear and non-nuclear.

I don't need to speak of the economic benefit of Vermont Yankee since many business leaders have already done so. Nor do I need to speak about the great benefit local charities receive, you've heard from them as well. You, as board members, are smart individuals, and I'm certain you recognize the public good Vermont Yankee provides in regard to those issues.

Instead, I would like to speak of my personal experiences working at different types of power generation facilities — the facilities that would most likely replace much of the energy generated by Vermont Yankee should it close.

I would like to tell you about a fossil plant I worked in last February where after just six weeks the paint on my car was damaged by the acid rain coming from the stack. The pollution that damaged my car does not know state borders and simply goes wherever the wind blows, including Vermont. This is certainly NOT in the public's good.

I would also like to mention a Biomass plant, a so-called "Green Plant," that I worked in a few years back. After just two shifts, the caustic chemicals used in the scrubbers ate up the threads in the soles of my shoes causing the soles to fall off. These chemicals were found on the walkways and catwalks and when it rained they simply went down the street and ultimately into the ground water which also doesn't recognize state borders. This is certainly NOT in the public's good.

I've worked at wind turbine projects in Iowa, Minnesota, Oklahoma, California, and Texas. At each of these sites I found dead birds that had been killed by the turbines. These birds did not know state borders either and some of them may have come from Vermont. Again, this hidden side effect of wind energy is certainly NOT in the public's good.

And what about the coal plant in southern Nevada with tens of acres of fly ash containing arsenic and God knows what other carcinogens? All in public view where the winds were free to carry it where they pleased. Most certainly NOT in the public's good.

So I close my testimony by asking each of you to consider the true environmental impact of these alternative sources of energy. I encourage you to personally visit some of these plants to see first-hand, as I have, the impact of these plants in regards to the air we breathe, the water we drink, and the wildlife we enjoy.

★★★★★

Stuart Endsley works at Vermont Yankee. He gave this statement at the interactive TV hearing and was kind enough to send Meredith a copy of his statement.

9. Global Warming and Vermont Yankee

By Ellen Cota

I am Ellen Cota. I am a mother of two active kids. I live in the Emergency Plan Zone, am an active community member, a board member and a consumer of local goods.

I am also a proud employee at VY. I work about ½ a football field away from the reactor. I am the Operating Experience Coordinator at Vermont Yankee and I am responsible for reviewing events and issues at other nuclear facilities so we can learn from them and implement best practices.

This is critical time in history. Like many, super storm Sandy has really frightened me. The reality of climate change is becoming more real with each storm. I am proud of the work at VY and know that this facility is not contributing to climate change. Nuclear energy is the only clean-air source of electricity generation that provides large amounts of power 24/7. We as a nation, not only VT, need to embrace nuclear.

The facts are that coal and natural gas power plants not only pollute more, they also have a lower capacity factor or amount of electricity produced compared to the maximum it could produce. Coal is 61%. Natural gas is 46%. Nuclear is around 90%. Nuclear is more reliable and cleaner. Vermont Yankee historically is one of the most reliable nuclear generating facilities of its type.

For my kids sake and the generations to come, please approve the Certificate of Public Good.

10. Avoid Carbon Dioxide

Keep Vermont Yankee

By Dr. Carlos Pinkham

On the evening of November 7, I spoke before the Vermont Public Service Board in Vernon in support of the continued operation of Vermont Yankee. What appears below is similar to the testimony I submitted in written form to the board, and from which I drew my remarks.

In the '70's, I was one of the U.S. Army's first environmental scientists working National Environmental Policy Act issues with Project Managers. I found that progress could be made if both parties would adopt four ground rules:

- Agree that both sides had legitimate concerns.
- Agree to listen seriously to these concerns and see the problem from the other side's perspective.
- Agree to think outside the box.
- Agree to be willing to compromise.

Employing these four ground rules, we were able to ensure completion of the mission and mitigation of environmental impacts in each of the hundred plus projects I worked on. I sense that some of these rules are not being followed in the VY controversy.

As Vermonters, we have always taken pride in being environmentally conscious. We keep our streets clean, we protect our parks and lands, we recycle, and we have some of the country's highest air quality ratings. These are the same sources of pride that motivate both sides of the Vermont Yankee debate.

Following, are some sources of pride that should be considered rationally from the pro Vermont Yankee side. Since coming online in 1972, Vermont Yankee has generated over 150 million megawatts of electricity and prevented more than 69 million tons of carbon dioxide from entering our air. Vermont Yankee's clean operation avoids 2.8 million tons of carbon dioxide emissions annually, as well as emissions of sulfur dioxide and nitrogen oxides. Avoidance of these harmful greenhouse gases while producing almost three-quarters of the electricity generated in state has helped Vermont achieve its second-place nationwide ranking for lowest electric power-related carbon footprint.

In addition, Vermont Yankee has contributed significantly to Vermont's number one ranking for "green" jobs per capita, according to a March, 2012 report by the U.S. Bureau of Labor Statistics. Nuclear power jobs, including hundreds at Vermont Yankee, are clearly defined in the report as green jobs.

We would lose these sources of pride with a Vermont Yankee shutdown. MIT Professor and head of the Dept. of Nuclear Science Dr. Richard Lester notes that Vermont Yankee's closure would lead to significant increases in carbon emissions. If the 620 MW of power provided by the plant is replaced by natural gas-fired plants on the New England grid, the increase would be equivalent to 30% of the state's total current emissions.

Vermont has always been an environmental leader. If we choose to close down Vermont Yankee, our leadership could suffer, as closing Vermont Yankee will lead to heightened use of fossil fuels to meet electricity demand.

Carlos F. A. Pinkham, PhD
Active Professor Emeritus

Co-Director, Vermont State Science and Mathematics Fair
Chair, Vermont ESTEEM Network
Researcher, Vermont EPSCoR Streams Project
Faculty Advisor, Norwich Christian Fellowship

★★★★★

Norwich University, founded in 1819, is the oldest senior military college in the United States.

Safe and Reliable

11. Vermont Yankee Is an Asset to Vermont

And the Sky Is Not Falling

By Deb Schulze

My name is Deb Schulze. I am a Vermont native and live in North Springfield. I am speaking tonight in favor of granting Vermont Yankee a Certificate of Public Good.

Vermont Yankee provides safe, clean and reliable electric power.

I have observed with increasing skepticism the tactics of VY's opponents. These folks are trying to scare everyone into believing this plant is old and broken down, and that just is not true.

VY is like most any business, it is continually maintained and upgraded and it is also carefully monitored by the Nuclear Regulatory Commission.

Vermont Yankee has saved more than 50 million tons of carbon and other pollutants from being released into the environment.

We now live in a world of big plasma televisions, electric cars and a whole host of electronic gadgets. Our need for electricity is continually increasing. Closing down Vermont's largest producer of electricity (two-thirds of the in-state generation) makes absolutely no sense.

Vermont Yankee, along with and its employees and contractors provides more than $100 million per year, into both the local and state economy. Taxes and payments to the state account for nearly $15 million.

In 2011, Entergy Vermont Yankee contributed approximately $435 thousand to the community through Entergy open grants, the Entergy Charitable Foundation, site sponsorships, and annual events. This amount includes Vermont Yankee employee pledges/ company match to the tune of $109,000 to the United Way.

The sky is not falling. Through all of the negativism generated by the anti-nuke activists, Vermont Yankee has been quietly and continuously producing safe, clean, reliable power for over 40 years.

I hope you will grant it a certificate of public good so that it can continue to do so.

★★★★★

Deb Schulze's husband, Bill, works at Vermont Yankee. She was kind enough to share her statement and provide this picture.

12. Phobias Should Not Determine Policy

By Peter Roth

My name is Peter H. Roth and I am a resident of the Town of Hartford Vermont. I am here this evening to petition this board to support and issue a Certificate of Public Good for the Vermont Yankee Nuclear Power plant.

I have thought long and hard about this issue, and as an experienced chemical engineer and scientist, with a strong interest in safe, reliable and inexpensive electric energy, I have concluded that the Yankee plant should remain open for another 20 years. My arguments are short and reasonable to support my position and I want to share them with this board.

There is no rational argument to shut a facility that continues to produce safe, reliable and low cost electricity for Vermont and the New England grid and has demonstrated so for a long time. Electricity is not a luxury, but a vital necessity as we know, when a storm like Hurricane Sandy shuts down power for millions.

Loss of power creates a high level "Misery Index" for people, but creating a condition that raises electricity costs for marginal income folks, also creates a "Misery Index." We are dealing with a commodity that is essential to our lives and an option that cannot be deferred. Food, clothing, shelter and power cannot be deferred. If electric rates go up, the only option is to cut another expense. This impacts people at the lower end of the economic scale and increases their "Misery Index."

Those that fear "Nuclear Power" may suffer from the same anxiety and condition that causes fear of flying, or fear of heights, or fear of enclosed space and no rational argument can dissuade them from their phobias. However their fear should not be an argument that impacts the lives of more rational people.

Many say: substitute nuclear power with wind, solar, Quebec Hydro, or ISO New England grid. However Quebec Hydro and ISO-NE grid also depend on nuclear power as part of their grid contribution. The electrons we buy from the grid don't know where they came from, whether Seabrook or Pilgrim or Gentilly 1 & 2. Since we buy most of our electricity in Vermont from Quebec Hydro and ISO-New England grid, we are "De Facto" supporting nuclear power that someone else has produced. The only honest course, if you don't like nuclear power, is don't buy from the grid and become self supporting and see how long you can support such an energy cost.

The final argument for maintaining Vermont Yankee, is jobs and the economy. The Vernon plant supports hundreds of good paying jobs and contributes a large tax revenue into the Vt. economy. Try to look someone in the face, that support family and children, and tell them, I was responsible for your loss of job and income. Then try to justify what you did that weakened the local and state economy, upon which so many depend.

In summary, the issuance of a Certificate of Public Good is a very logical and reasonable action that this board is empowered to do. The Vt. legislature created this board to act as a non-political but expert entity to oversee important public energy issues. Please act in support of your neighbors and citizens.

★★★★★

Peter Roth has a BS and MS in Chemical Engineering, and wide engineering experience ranging from missile systems to imaging. Roth holds twenty-two U S patents, with more patents pending. He is a founding member of the Coalition for Energy Solutions, and you can read more about his background at the Coalition website. Roth leads classes at Dartmouth's ILEAD. In the fall quarter, he led a course "Save the F E W (Food, Energy, Water)".

13. A Safe Plant and the Same Few Protesters

By Steve Moriarty

Hello. My name is Steve Moriarty; I live in Greenfield, MA and have worked at VY for 34+ years.

Prior to that, I worked all around NE for New England Power Service Co., a construction company that services all NE plants.

My background is electrical construction/maintenance, Materials Management, Quality Assurance and Quality Control.

In that first job, I traveled from fossil plant to fossil plant in NE. Some were oil and some were coal. I also work at a few Hydro stations.

In 1977 I was assigned to VY. When I first visited the site I was pleasantly surprised. It was clean, organized and run proficiently.

It was nothing like the dirty fossil plants that I was working in previously.

I soon saw that extraordinary attention was and continues to be paid to safety of the public, workers, and environment.

Procedure adherence was not an option; it was and continues to be the standard operating procedure.

I was so impressed by the workplace and the personnel at VY that I took a permanent position at VY and have been here since (now 39+ years combined service).

I met and married my wife, raised two children in Franklin County.

The bulk of my state taxes are paid to VT.

My daughter worked here as a summer employee while attending college. She is a Microbiologist and is now married with two healthy and happy children.

My son is a Mechanical Engineer. He worked here for two summers as an intern, and that experience set him up for success. He too, is now married and healthy.

I would not have allowed either to work here, if I didn't truly believe it was a good and safe place to work.

Both of my children and my wife cannot believe the day to day difficulty that I must go through as a nuclear worker just to be able to do my job.

This defensive posture, that all nuclear workers are forced to assume, is not a daily task of workers in most professions.

My son noted that "there really aren't many young people speaking out against your plant." This was based on discussions over the nuclear option at UMASS Amherst, while at college.

I got to thinking, and I too realized that I have been looking at the same faces for 10-20-30 years. (i.e. really not a lot of people speaking out against Nuclear, "just the same few").

I also asked my son what he has heard at college and in the workplace, relative to nuclear power and its future, and his response was that many students in the Engineering program and engineering professionals are in favor of nuclear as a viable green source of energy.

Some, of course, like the idea of wind, solar, or hydro; but they also realize the nuclear power is needed, as well.

One thing that was agreed on was that this country's reliance on coal and oil was something they don't want.

They all seem to be very concerned about the environmental damage being done by fossil fuel, let alone the problems associated with obtaining those fuels.

So, what's the point? The point is that if "these same few" spent as much energy on working with us to make the next generation of nuclear power plants even better, we all could gain.

It may be time to stop the 10-20-30++++ years of protesting and redirect that energy toward valuable and welcomed input to the future, all of our futures.

Nuclear power is here, it has proven its value, and it can be even more valuable to the environment and our energy demands into the future.

Thank you for your consideration, and I hope you see the real value of allowing Entergy-Vermont Yankee to operate into the future.

★★★★★

Steve Moriarty submitted this statement to the Public Service Board by email. He was kind enough to send Meredith a copy for posting.

Moriarty's and Stuart Endsley's posts are boots-on-the-ground comparisons of nuclear and fossil plants. Their words may be well known to utility workers, but they cover subjects that you rarely see in the media.

14. Vermont's Fair Share of the Grid

By Howard Shaffer

Operating Vermont Yankee is Vermont's "Fair Share" of the Grid

Operating Vermont Yankee is in the Public Good because it is Vermont's "Fair Share" of the New England grid. Vermont depends on the grid and also supports the grid. The grid is a shared resource and commitment, between all the members. For the overall good, all members must accept things that may not be optimum for them, or things that they don't agree with, or like.

Vermont depends on the grid to supply its out-of-state power, and to buy its excess power.

Vermont Yankee (VY) is not currently selling its power to Vermont, but this is not all it contributes. VY provides capacity to the grid that does not emit CO2. Without VY, gas plants would run to provide the difference, in the short and medium term. The New England carbon footprint would be bigger without VY. This diminished CO2 emission is part of Vermont's contribution to the region.

Without VY the grid will be less robust than it is with the plant. The grid operator has said it will be adequate without VY, but with it the grid will be better, by having more capacity available for emergency situations.

The VY plant provides reactive power services to the grid at its location. Without the plant the compensation would have to be provided from other sources or equipment.

The items to be considered for the Public Good speak of out of state projects that benefit Vermont, but not the opposite situation. The items don't seem to address the situation of a project in Vermont that does not sell power to the state, yet supports the whole grid, and provides other interstate benefits.

The concept of "Fair Share" may be illustrated by asking Vermonters if they still want gasoline for their vehicles. If they say "Yes" then ask them if they will accept a refinery in the state. If they oppose a refinery, yet still want gas, ask them how it is fair that some other location and people bear the burden of the refinery for them.

Respectfully submitted
Howard Shaffer, PE, Vermont license 5419

★★★★★

Howard Shaffer was a startup engineer and a systems engineer at Vermont Yankee and several other power plants. He is a long-time member of the American Nuclear Society, and was an ANS Congressional Fellow in 2001. Recently, he has led the ANS Vermont Grassroots project. Read more about Shaffer's background in "Monday Blue Ribbon, Howard Shaffer and Public Outreach" at the blog *Yes Vermont Yankee.*

Shaffer works closely with Meredith Angwin on energy education and pro-nuclear advocacy. Both received ANS Presidential Citations in 2012. Shaffer blogs monthly at *ANS Nuclear Cafe,* and writes frequent guest posts on the blog *Yes Vermont Yankee.*

Supports the Community

15. Vital for the Region and My Family

A Teen-Agers View of Vermont Yankee

By Evan Twarog

My name is Evan Twarog. I am a sophomore at Keene High School, and have lived in this region for more than 13 years while my dad has worked at Vermont Yankee. In that time, I have grown to love the region, and couldn't have imagined growing up anywhere else. Each year, I play two school sports, and I am a proud member of the Keene Rotary Interact Club.

As I've seen the legal battle pan out over my family's primary income source, I've learned just how vital Vermont Yankee is to our community. Each year, the nuclear energy facility provides local charities and non-profits with hundreds of thousands of dollars in funding which helps ensure mouths go fed, athletes have gear to play in, and disabled citizens are taken care of. My father is the president of one of those organizations. For almost 14 years, he has worked at the facility and has risen to the position of Shift Manager. In the time he's worked there, my confidence in Entergy and its employees has only increased.

At the facility, there is no higher priority than safety. In fact, he's spent more time achieving and maintaining his reactor license than he spent in the four years it took him to get his Bachelor's. In the past couple of years, as my interest in nuclear energy and VY has increased, I have delved into issues concerning the plant such as economics, safety and environmental issues, and this research has only made me more certain just how important VY is to Vermont's economy and environment.

In these rough economic times, it would be an enormous mistake to try to shut down Vermont Yankee and force the 600 families supported by the facility to move out the region. We have much to gain by keeping the facility operating, and much to lose by shutting it down. Thank you.

★★★★★

A short note about the picture. At the Vernon hearing, people were asked to line up so that they could get to the microphone quickly. In this picture, you can see Cheryl Twarog standing just behind Evan, ready for her chance to speak. Between them, sitting down, is Peter Lothes, who also spoke for the plant. The man in a robe, also in an aisle seat, identified himself as Brother Toby from a Buddhist temple. This man was quoted in the *Brattleboro Reformer*. He spoke against the plant and against all nuclear energy.

16. Over 600 Families

By Cheryl Twarog

My name is Cheryl Twarog, and I'm proud to say that my husband works at Vermont Yankee. At Vermont Yankee there are 620 of the most dedicated, hard-working, highly trained individuals you'll ever meet. But there's another chapter in this story that often goes untold. That's the story of the Vermont Yankee families: the moms & dads, husbands & wives, sons & daughters of the employees.

At Vermont Yankee we are 620 families strong.

- That's 620 families that count on the affordable health insurance that Vermont Yankee provides.
- That's 620 families that rely on the $65 million in wages paid each year.
- That's 620 families that worry that we will have to put our homes on the market, find another job in this tough economy, or move our children from the only schools they've ever known.
- And that's 620 families that are holding our breath hoping that you will agree that it's for the public good that Vermont Yankee be given a certificate for continued operations.

My family is one of the Vermont Yankee families, and we realize just how much we have to lose if Vermont Yankee closes prematurely. My husband, John, has worked and trained at Vermont Yankee for almost 14 years. His skills are highly specialized and wouldn't easily translate to other jobs in the area, making it necessary to pack up our family and move. There is much that we would miss:

Such as the way the lives of everyone in our family have been enriched by our time spent volunteering for local non-profit organizations.

Or the many opportunities our boys have been blessed with, many of which are unique to the area…

- Like the Keene State Orchestra and Keene Chamber Orchestra welcoming in our 12 year old son, Cam, who lives and breathes the violin.
- Or the highly regarded service organization that has allowed our 15 year old son, Evan, to develop and demonstrate leadership skills, along with his compassion for others.

Both boys have ambitious goals for their futures.

The simple truth is that much of this is possible because of Vermont Yankee, and our lives would look dramatically different if VY were to close prematurely. As you work to make your decision I hope you'll remember that there are 620 families just like ours, who depend on the continued operation of Vermont Yankee.

Thank you for being here tonight and giving us the chance for our voices to be heard.

★★★★★

At the Vernon hearing, Cheryl spoke directly after Evan spoke, and started her statement with a cheerful: "That's my boy!"

Cheryl Twarog and her family made these buttons. You could see many people wearing them at the hearing.

17. The Ability to Live in My Home

By Karen Wilson

Good evening, my name is Karen Wilson.

I am here to support Vermont Yankee in obtaining the Certificate of Public Good.

I live here in Vernon with my daughter Heather. Who happens to be an adult with developmental challenges.

My other daughter, Amanda also lives here in Vernon with her partner, Jason and my two grand-daughters, Kali and Reis.

I moved to the area in 1971 and began raising my family here in 1980.

I worked at a local business in Brattleboro until a few years ago, when due to the times, found myself in a position like many and was laid off.

Thankfully, just over a year and a half ago, I was offered a job and accepted a position at Vermont Yankee.

Vermont Yankee has many programs and offers support not only to the community but to its employees. With the support of management and my fellow employees at Vermont Yankee, I am able to take advantage of one program they offer, that is allowing me the opportunity to go back to school to complete my business degree.

Having Vermont Yankee here in Vernon, as an employer, has made it possible for Heather and I to continue to live in our home, for me to support my family and for me to continue my education.

I personally feel, and have experienced firsthand, that Vermont Yankee is a safe and vital contributor to its employees and to the community.

Thank you for the opportunity to share my story with you.

18. Buy Local and Help Your Community

By Kenyon Webber

My name is Kenyon Webber and I moved here to work as a Design Engineer with Vermont Yankee. I am here to bring three points to your attention.

First, about the economy of the area. A few months ago, I had the privilege of attending the Regional meeting of the group SEVEDS, South Eastern Vermont Economic Development Strategies. The findings of this group were quite disturbing, but not surprising to me. They found that the average income for the state is $38,800 and for the southeastern region it is around $32,500 with a college degree. The national median income is $54,756, according to the Dept. of Labor Bureau and Statistics. SEVEDS also reported that there is a decline in the population in the southeastern region of Vermont of people age 44 and under. Those of us who moved here to work for this facility would like to continue to work here. Also, the area would lose over $6 million in taxes that Vermont Yankee presently provides to this community.

Second, Vermont Yankee is a good community partner. I personally have worked with three very successful projects in the two years I have been with the company: Project Feed the Thousands, work on the Boys and Girls Club of Brattleboro, and the remodeling of the Brattleboro Drop-In Center. Without the work of these dedicated people from the facility, many people would not be fed and the Boys and Girls Club and Brattleboro Drop In Center would still be in major disrepair.

Third, this area should be committed to the "buy local" motto. I suppose many of you feel that because this is not some farm stand on the side of the road, it is not a local business. This business employs hundreds of local people that support the farm stands and other local businesses. We live here locally and spend our money locally, just like any other person in this community.

So, I close with three good reasons to vote favorably for Vermont Yankee. We provide higher wage, stable careers for more than 600 people, not to mention the millions of tax dollars we provide; we are a good community partner, and you should be buying your electricity local.

★★★★★

Kenyon Webber has an electrical engineering degree from Tennessee Technological University. She and another young engineer, Brian O'Callahan, were featured on the Vermont Yankee website during National Engineers Week.

19. Vermont Tourism Supported by Vermont Yankee

By Heather Sheppard

For years I have been working on the front lines of the Vermont tourism industry, and it is from this vantage point that my support for Vermont Yankee has developed. On Nov. 7, I traveled to Vernon – a six hour drive round-trip – to tell the Vermont Public Service Board why I support Vermont Yankee.

For many years, I have worked directly with the guests at one of America's leading family destination resorts. They tell me they come to Vermont because it is beautiful and unspoiled, because the air is clean and the views are gorgeous. If they wanted dirty air or visual clutter, they would have gone somewhere else, or just stayed home. Most of my guests come from nearby out of state - from the other New England states, New York, or the two closest Canadian provinces. Most of them drive part or all of the way. In short, they come not only for the fun of the resort itself, but for the beauty of surrounding Vermont, especially the mountains, for the clean air, and because its nearness makes it affordable, compared to the resort competition out west.

Beauty, clean air, and affordability - Vermont Yankee is a benefit to all three. Beauty, because having an operational Vermont Yankee means we are in less of a rush to clear cut our mountain ridgelines and valleys to make way for wind farms and for crisscrossing new power lines for the hodgepodge of small-scale power generation that some would have replace it. Clean air, because Vermont Yankee emits no air pollutants, unlike the coal and gas plants that will be ramped up if Vermont Yankee closes. Some environmental groups that should know better have suggested a patchwork quilt of woodburning power plants, carbon emissions and all, to replace Vermont Yankee. From an air quality point of view, this makes no sense. To me, one of Vermont Yankee's greatest environmental benefits as a power producer is that it already exists. No more trees need to be cut down, nor rocks blasted, nor tourist-drawing scenic views destroyed. There is no need for lines of slow, loud, exhaust-emitting trucks running to and from construction sites and woodchip plants.

Finally, Vermont Yankee power is very affordable, especially when compared to the renewable power that Montpelier seems determined to make us pay for. Unfortunately the paying public isn't just my family. Resorts use a lot of electricity, and it is a simple rule of business that the customer always pays in the end.

Heather Sheppard
North Cambridge, VT

20. Young Workers in Windham County

By Lindsay Rose

Good evening commissioners,

My name is Lindsay Rose, and I would like to first thank you for giving me the opportunity to speak to you this evening. I'm from Hinsdale, New Hampshire and graduated from UNH in 2010. About a year after I graduated I got a job at Vermont Yankee where I found an exciting opportunity, a well-paying job, and a company with good community values.

It's a fact that the demographics in Windham County are changing. From 2000 to 2010, the number of older workers, or the baby boomer generation, increased, but the number of young workers, aged 25 to 44, declined. Basically Windham County's future workforce is leaving.

Vermont's future workforce will be looking for opportunity: A steady job, a well-paying job, a company that gives back to their community, and an industry that's contributing to the overall good and health of our country. VY provides these jobs: more than 600 of them. It was ranked among the Best Places to Work for the last three years in a row by Vermont Business Magazine; VY management encourages volunteerism in our communities; and the nuclear energy industry is contributing to a clean, safe, reliable energy future.

Vermont Yankee is a company that brings professionals to Vermont. These professionals include recent graduates like me, and our country's military veterans with nuclear experience who are ready to bring their skills and experiences to Vermont. Vermont Yankee is an essential economic driver and is a business that provides career opportunities that keep people in Vermont.

At Vermont Yankee I have a steady, well-paying job, I'm encouraged to learn and take on new activities and experiences, and I'm given opportunities to help the community by participating in VY's many volunteer activities. This year I joined my co-workers and helped remodel the Brattleboro Drop In Center, I'm supporting Project Feed the Thousands, and I have begun volunteering for Big Brothers Big Sisters.

I've chosen to stay in the area specifically because of my job at Vermont Yankee and I enjoy living in the area. The opportunities that I've described are what keep me working at VY, and are what keep me in the area. If I didn't have these opportunities at Vermont Yankee, I would find them elsewhere. This is a great place to live if you have a good job, but I couldn't live here without one. Thank you for considering my comments.

★★★★★

Lindsay Rose spoke at the Vernon hearing on behalf of herself, as a personal statement.

21. Vermont People Work at Vermont Yankee

by Patty O'Donnell

Patty O'Donnell spoke from notes, not a written statement. She stated that Vermont Yankee is a safe plant, and one of the top-rated in the country. Workers wouldn't live near the plant and send their children to school near the plant if the plant were not safe.

O'Donnell also spoke against the demonizing of "Entergy Louisiana." Part of her statement was quoted in the *Brattleboro Reformer*:

> *Patty O'Donnell, chairwoman of the Vernon Selectboard and former legislator, said Yankee is "a high-quality nuclear power plant ... one of the best in the country."*
>
> *"This isn't Entergy Louisiana," said O'Donnell. "It's Entergy Vermont Yankee. Look around this room you'll see the faces of the people who run this plant. That's Vermont Yankee."*

Appendix

The table contains a list of the statements included in this book, along with the hearing at which each was given and the date posted on the blog *Yes Vermont Yankee.*

#	Name	Hearing	Date Posted
1	Charles Kelly	Vernon	November 17, 2012
2	Peter Lothes	Vernon	November 18, 2012
3	Richard Trudell	Vernon	November 14, 2012
4	Dianne Amme	Vernon	November 17, 2012
5	Bruce Shields	Vernon	November 9, 2012
6	Bill Schulze	Vernon	November 10, 2012
7	Meredith Angwin	Vernon	November 11, 2012
8	Stuart Endsley	interactive TV	December 17, 2012
9	Ellen Cota	Vernon	November 16, 2012
10	Carlos Pinkham	Vernon	November 15, 2012
11	Deb Schulze	interactive TV	January 4, 2013
12	Peter Roth	Vernon	November 11, 2012
13	Steve Moriarty	email	December 23, 2012
14	Howard Shaffer	Vernon	November 10, 2012
15	Evan Twarog	Vernon	November 18, 2012
16	Cheryl Twarog	Vernon	November 18, 2012
17	Karen Wilson	Vernon	November 14, 2012
18	Kenyon Webber	Vernon	November 12, 2012
19	Heather Sheppard	Vernon	November 13, 2012
20	Lindsay Rose	Vernon	November 9, 2012
21	Patty O'Donnell	Vernon	November 16, 2012